T0294378

Published by Ice House Books

TM © 2020 IFLSCIENCE Limited. All rights reserved.

Written and Designed by Smart Design Studio

Photography Shutterstock.com – for individual credits see back page

Ice House Books is an imprint of Half Moon Bay Limited
The Ice House, 124 Walcot Street, Bath, BA1 5BG
www.icehousebooks.co.uk

The material in this publication is of the nature of general comment
only, and does not represent professional advice.

ISBN 978-1-912867-64-6

Printed in China

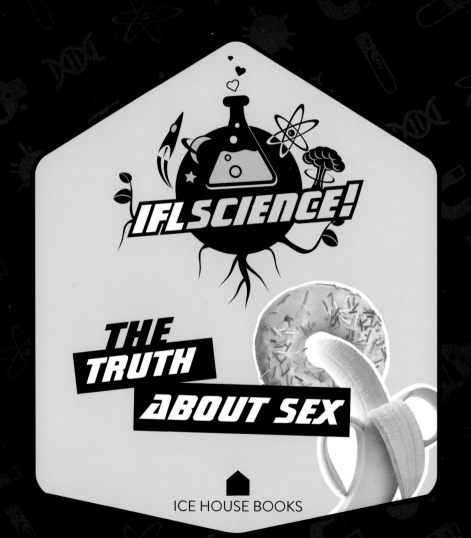

IFLSCIENCE!

THE TRUTH ABOUT SEX

ICE HOUSE BOOKS

CHEMICAL ATTRACTION

WITHOUT SCIENCE THERE WOULD BE NO ORGASMS!

Both the clitoris and the head of the penis contain a high number of nerve endings. When stimulated they send signals to the brain, which results in the release of large amounts of happy hormones, **dopamine** and **oxytocin**.

These chemicals give us feelings of **pleasure** and **intimacy** when we orgasm.

SCIENCE MAKES US HORNY!

BIG FEET = BIG PENIS

DOES SHOE SIZE REALLY MATTER?

NO. SORRY, THIS IS A MYTH.

There's no scientific link between shoe size and penis size. The same goes for a man's height.

BIG SHOES, SO WHAT?

CALORIES BURNED DURING SEX

CAN DOING IT REALLY HELP YOU LOSE WEIGHT?

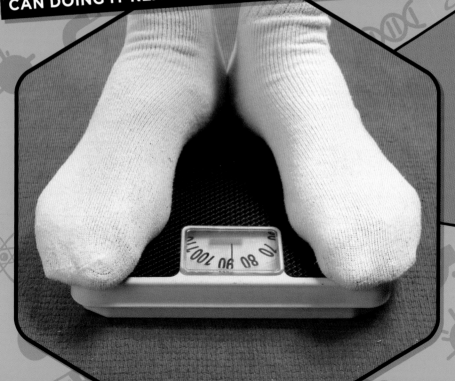

YES!

But there's no hard scientific evidence to prove just **how many** calories a steamy sex session could burn off ... a recent study claims it's as little as 21 per romp!

Calories burned depends on variable factors such as **position**, **speed**, **length of session** etc.

THRUST COUNT

HAVE YOU EVER WONDERED WHAT THE AVERAGE 'THRUST COUNT' IS PER SEX SESSION?

NO? JUST US THEN...?

Well, in case you **WERE** wondering how many thrusts a man averages between the sheets, here's your answer – **between 100 and 500 thrusts!**

THE AVERAGE TIME OF INTERCOURSE IS APPROX. 17MIN 5SEC.

17
MINUTES

GRAVITY = CONTRACEPTION

A GIRL CAN'T GET PREGNANT WHEN SHE'S ON TOP...
CAN SHE?

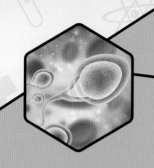

OH YES, SHE TOTALLY CAN!

Whether you're rocking it **doggy style**, **missionary**, **reverse cowgirl** or **girl on top**, a girl can get pregnant in **ANY** position if you have unprotected sex.

SPERM ARE STRONG SWIMMERS!

¡69
The upside down

According to the survey of stress-inducing sex positions, **anal sex** and the '**kneeling wheelbarrow**' came in at **2nd** and **3rd place**!

THE BLOOD IS RUSHING TO OUR HEAD JUST THINKING ABOUT IT...

17

CAN MASTURBATION...

...REALLY MAKE YOU BLIND?

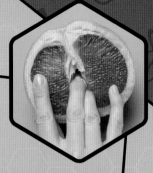

CHILL OUT – *NO.* OF COURSE IT CAN'T.

It also doesn't make you **grow hair on the palms of your hands**. Or lead to **impotence** later in life. Or cause **mental illness**. Or take away your **virginity**.

FACT: ALL BODY PARTS PERFORM BETTER IF USED OR EXERCISED REGULARLY, AND OUR SEXUAL 'PARTS' ARE NO DIFFERENT.

DO IT YOUR SELF!!

WHAT TYPE OF MEN...
...GIVE THE BEST ORGASMS?

According to research, men who produce higher orgasm rates in women possess certain positive character traits.

Humour, **creativity**, **warmth** and **faithfulness** are all box-ticking traits when it comes to looking for a partner who'll keep you happy between the sheets.

THE SAME STUDY ALSO CLAIMS THAT PARTNERS WHO SMELL GOOD GIVE US BETTER ORGASMS.

SMELLING IS SEXY...

A STRONG SENSE OF SMELL CAN LEAD TO BETTER ORGASMS!

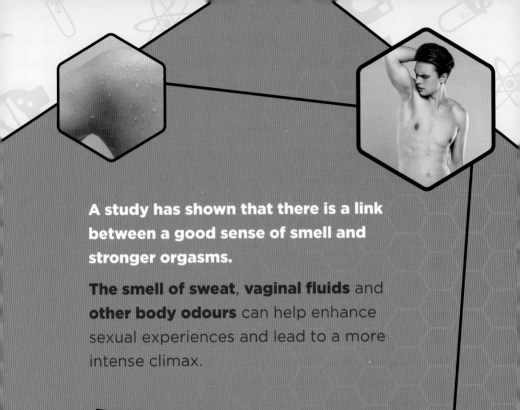

A study has shown that there is a link between a good sense of smell and stronger orgasms.

The smell of sweat, vaginal fluids and other body odours can help enhance sexual experiences and lead to a more intense climax.

SNIFF, SNIFF!

THE KEY TO A HAPPY SEX LIFE

...IS SOMETHING VERY UNSEXY, ACCORDING TO SCIENCE!

CONSCIENTIOUSNESS.

YEP, YOU READ THAT CORRECTLY.

According to a new scientific study, people who are **conscientious**, **organised** and **great at planning ahead** are more satisfied in their sex lives than those who aren't.

GET PLANNING!

WET DREAMS

...AREN'T JUST FOR MEN!

Although ejaculation during sleep is much more common for men, studies show that up to **40% of women** have experienced '**wet dreams**' in the last year.

SWEET DREAMS!

27

SIZE = SATISFACTION

SORRY, LADS, ANOTHER SEX MYTH BUSTED BY SCIENCE!

BUSTED

A woman's '**G-Spot**' is located approximately two inches inside the vagina. **This means that large penises often completely miss the sacred spot when thrusting!**

IT'S WHAT YOU DO WITH IT THAT COUNTS!

FEMALE EJACULATION

HAS THE INTERNET BEEN LYING TO US?

YES.

Research shows that 'squirting' during climax is actually **involuntary urination**.

Think about that next time you search for 'squirting' on the internet....

FACT: SOME WOMEN CAN 'SQUIRT' UP TO 5ML OF FLUID!

HOW OFTEN DO MEN THINK ABOUT SEX?

IS IT REALLY EVERY SEVEN SECONDS?

NOT QUITE...

A university study found that on average a man thinks about **sex 19 times a day**.

By comparison, they think about **food 18 times a day**, and **sleep 11 times**.

ONE-TRACK MIND?

THE HOLY TRINITY OF THE FEMALE ORGASM

THE SCIENTIFIC ONE, TWO, THREE TO KEEP HER COMING BACK FOR MORE...

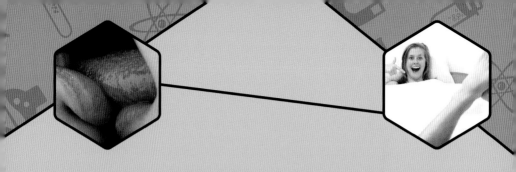

When it comes to making a lady happy, scientists have done the groundwork for you.

A recent scientific study reports that only 35% of women claim to orgasm through penetrative sex alone. However, **up to 80% of women reached climax after this golden trio was applied:**

1. **Deep kissing**   *2.* **Genital stimulation**   *3.* **Oral sex**

DON'T THANK US, THANK SCIENCE!

35

WORLD CUP W**K

A WELL-KNOWN PORN WEBSITE RECENTLY PUBLISHED A TREASURE TROVE OF DATA...

...WHICH INCLUDED POPULAR SEARCHES DURING THE 2018 WORLD CUP.

According to the sexy site's data analysts, '**football**' and '**soccer**' were popular searches during the tournament.

It would seem the site's users were also rather distracted whilst the football was on. **Average site traffic dropped by up to 47%!**

HE SHOOTS, HE SCORES!

ORAL SEX

...IS NOT JUST FOR HUMANS!

Bobbing for apples ... charming the snake ... sampling the sausage ... when it comes to oral sex, it's NOT JUST US HUMANS who are getting in on the action.

Wolves, **bears** and **bats** have all been observed doing the deed.

COME UP FOR AIR!

NEVER TOO OLD TO TANGO

GRANDPARENTS GET HORNY TOO!

Despite younger generations pretending that their parents only had sex to procreate and their grandparents haven't '**done it**' in years, studies show that this simply isn't the case.

Approximately one third of women in their 80s are still having sex with their partners. So long as you're in good health, there's no reason to stop enjoying sex.

LOVE COMES IN ALL SHAPES, SIZES AND AGES!

Love

APHRODISIAC FOOD

CAN CHOCOLATE REALLY GET US 'IN THE MOOD'?

According to science, certain foods really can make us horny. Some obvious foods, like dark chocolate, will do the trick, but here are some more unlikely feel-good foods:

Garlic – contains the mineral chromium, which helps regulate our serotonin levels.

Mushrooms – rich in selenium and vitamin D, which help our moods.

Carrots – as well as high vitamin C content, they also contain the antioxidant beta-carotene, which keeps our brain healthy.

ARE YOU GETTING YOUR FIVE-A-DAY?

FASTER THAN BOLT

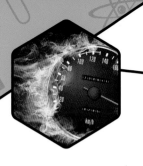

The initial jet of a man's ejaculation travels at approximately **28 mph.**

To put that in perspective, that's **faster than the current 100 metre world record**, which is **22.9 mph**.

READY, STEADY, GO!

SPERM COUNT

HOW MANY SWIMMERS DOES A MAN HAVE?

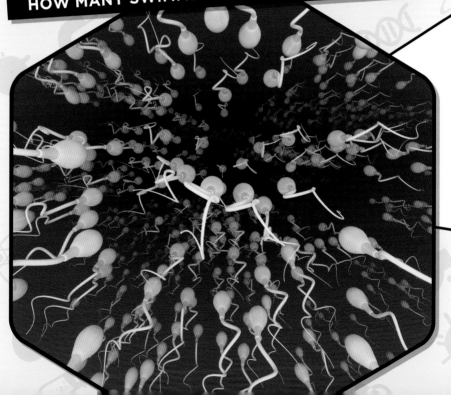

On average, a man produces **1000 sperm cells every second**.

THAT'S 86 MILLION A DAY.

And if you do the maths ... a single man has enough sperm inside him to impregnate every fertile woman alive!

ONE ... TWO ... THREE ... HOW MANY?

AS HAPPY AS A PIG IN...

...MUD, SO LONG AS THE MUD IS WHERE THEY ARE HAVING SEX!

CRAZY FACT ALERT...

A pig's orgasm can last for up to **30 MINUTES**!

Pigs are also **highly intelligent animals** who communicate with one another and like to sleep nose-to-nose. They dream when they sleep too.

OINK, OINK!

HOW OFTEN DO WE 'DO IT'?

ON AVERAGE, HOW OFTEN DO HAPPY COUPLES HAVE SEX?

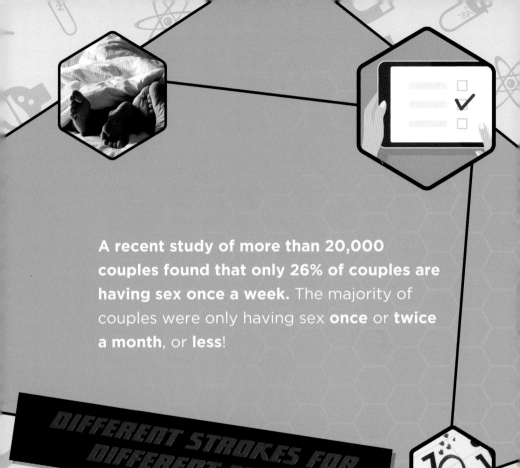

A recent study of more than 20,000 couples found that only 26% of couples are having sex once a week. The majority of couples were only having sex **once** or **twice a month**, or **less**!

DIFFERENT STROKES FOR DIFFERENT FOLKS.

CAN ORAL SEX GIVE YOU CANCER?

WE HAVE SOME BAD NEWS, GIRLS...

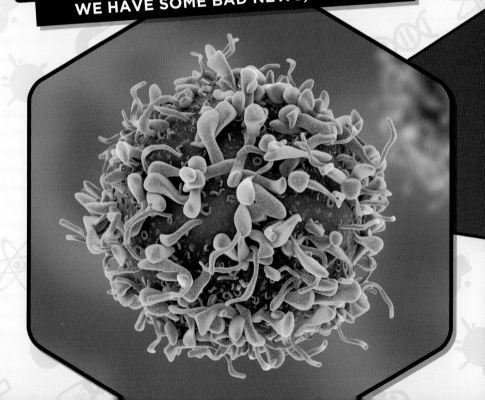

A recent study has shown that men who have a **high number of oral sexual partners** have an increased risk of head and neck cancer.

The reason? If a woman has HPV and a man goes down on her, he risks catching the infection. **The risk is higher if the man smokes.**

FACT: ORAL HPV IS RARE IN MEN WHO HAVE HAD FIVE OR LESS ORAL PARTNERS.

GOLDEN AGE

AT WHAT AGE DO WE HAVE THE BEST SEX OF OUR LIVES?

NO, IT'S NOT YOUR 20s ... OR YOUR 30s ...

...BUT YOUR 40S.

46 is reportedly the '**golden age**'
for the best sex of our lives.

STUDIES SHOW THAT THE MORE SEX YOU HAVE, THE HAPPIER YOU ARE!

SPERM SURVIVAL

WOMEN CAN FALL PREGNANT DAYS AFTER HAVING SEX.

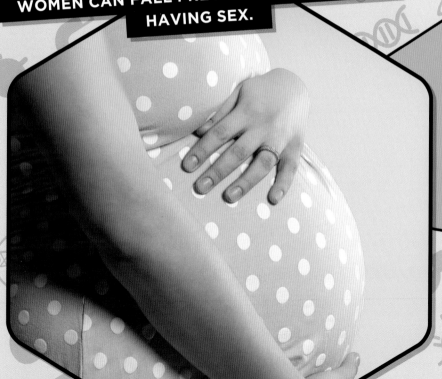

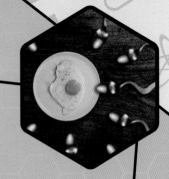

IT'S TRUE!

Sperm cells can live on in a woman's body for up to five days after intercourse.

A woman is fertile for approximately three to six days every cycle, which means that **IT'S POSSIBLE** to fall pregnant days after having unprotected sex.

STRONG SWIMMERS!

WHAT'S YOUR MAGIC NUMBER?

WHAT IS THE AVERAGE NUMBER OF SEXUAL PARTNERS?

A test group of sexually active adults between the ages of 20 and 59 were surveyed and asked what their '**magic number**' was.

THE ANSWER?

THE **WOMEN** HAD AN AVERAGE OF **FOUR** SEX PARTNERS DURING THEIR LIFETIME.

MEN HAD AN AVERAGE OF **SEVEN**.

17.5% OF MEN EXAGGERATE THEIR MAGIC NUMBER, COMPARED TO JUST 8.2% OF WOMEN.

SEX AND SPORT

CAN A ROMP BEFORE A BIG GAME AFFECT YOUR PERFORMANCE?

BIG SPORTING EVENT TOMORROW?

Go ahead, **wax your bean** … **peel your banana** … **give the old dog a bone**.

Studies show that sex before sport has no affect on performance at all.

BACK OF THE NET!

Photo Credits